NOTES FROM A MARINE

BIOLOGIST'S DAUGHTER

SAINT JULIAN PRESS

POETRY

Praise for NOTES FROM A MARINE BIOLOGIST'S DAUGHTER

Notes from a Marine Biologist's Daughter introduces the reader to a rich and varied lexicon of the natural world, its inhabitants, and the poet's interaction and close relationship with the same. These poems pay deep attention: to paddling in the Everglades, walking in the neighborhood, and indelible childhood memories. Nature is a never-ending source of remarkable occurrences for this poet-naturalist. We have new eyes for our world, reading this enthralled insightful collection.

—K. Alma Peterson
The Last Place I Lived
Was There No Interlude When Light Sprawled the Fen

What an extraordinary book! It struggles and accepts, it frets and studies, and then, pleased, relaxes into knowing. It is filled not only with Anne McCrary Sullivan's knowledge of botany but with loving tributes to her mother and grandmother, memories of her father, lost when she was a girl. These poems tremble the heart and, at the same time, teach the world -- of plants and flowers and trees. This book opens its arms to the world: "I am the tree, rooted and sturdy, I am the wind passing through." // "I am dust, I am desert, my banks overflow with flood."

Anne McCrary Sullivan is a naturalist by temperament and by training, and her responses to the world -- and to her foremothers — are large and generous and teach you more about the natural world than you knew you needed: "Our name is earth. Our name is sea. Our name is air." It is a pleasure to welcome this extraordinary poet's new work into the world.

—Deena Linett
When I Was Water
Translucent When Fired

Notes from a Marine Biologist's Daughter made me want to bow down and tap dance at the same time. It left me with a powerful (and lasting) feeling that, not only are we surrounded by quiet and not-so-quiet miracles, but we live inside and within a vast, complex, wonderfully inexplicable and explicable Being (called whatever you wish). There's no pantheism or fancy posturing here. Just the astounding beauty of reality. This work convinces me that poetry is not at the heart of everything. *It is the heart of everything.*

—Patricia Corbus
Finestra's Window
Winner of the 2015
Off the Grid Poetry Prize.

NOTES FROM A MARINE BIOLOGIST'S DAUGHTER

Poems

by

Anne McCrary Sullivan

SAINT JULIAN PRESS
HOUSTON

Published by
SAINT JULIAN PRESS, Inc.
2053 Cortlandt, Suite 200
Houston, Texas 77008

www.saintjulianpress.com

ISBN-13: 978-1-955194-14-3
Library of Congress Control Number: 2023933195

Cover Art Credit: Anne McCrary Sullivan
Author Photograph by Sara Fleming

In memory of
my mother, Anne Bowden McCrary,
a marine biologist when mothers weren't;
it has made all the difference.

CONTENTS

NOTES FROM A MARINE

BIOLOGIST'S DAUGHTER

PROLOGUE

EPILOGUE

Lexicon

I want to defy what I've been taught

fill my lines with scientific names, creatures
I have known—*tunicates, ctenophores, mollusca.*

And if I were to praise the moon?

In my childhood, Latin names were as ordinary as rice, as likely
to come up at the kitchen table—*bryozoans, renilla,* butter, please.

And that moon—over water, over mud. Always a prediction—
when to look for *limulus,* when to dig for *chaetopterus.*

Every word is ordinary   if you enter into relation with it.
No word is ordinary   if you enter into relation with it.

Copepod holds the memory of a mother   as wondrously as
bread.

I

When grief opens a cavern in me

I let the thunder roll there, echo
with the hawk's cry, the jay's shriek

until the blue-black sky cracks
and a rain comes heavy falling

until the laboring pines whorl
against the agitated oaks

until the earth grows dark
with wet, until the aquifer fills.

Fog

Channel and shores have disappeared.
Mists swirl, hover over invisible water.
I wrap myself against the chill.

There's an island out there.
I know it, see it on charts,
remember it from long ago.

When this deep fog lifts,
when I can read the compass,
locate myself in landscape,
advance towards an emerging
shape, I will paddle there.

Till then I sit, rope tied loosely
to a mangrove root, feel the rocking.

The Microscope on the Kitchen Table

Sometimes she slid it aside. We ate in its presence.

Sometimes we ate at a different table.

Sometimes we had a picnic on the back steps.

Sometimes she bent over it all morning.

Sometimes she bent over it at night.

She looked at water in the petri dish,
saw things we didn't see, stirred the water
with a needle-fine instrument.

Sometimes she could not contain her delight.
That cute little copepod
looks like he's riding a unicycle!

Sometimes she tried to show us,
but I usually couldn't quite focus.

All the while she monitored
pots on the stove, pans in the oven.

She counted. She wrote the numbers in notebooks.
Notebooks lined up on the shelf.

It's odd, how nothing changes

You are in the same place doing the same things.
Coffee tastes good. The small blanket on your lap
is soft as you wait for the sun to rise, glancing up
from time to time for signs—a patch of almost-light,
a silhouette of trees—then go on reading in the way
of every morning. Nothing has changed.
Except this one small piece of knowledge that floats
randomly in the back of your mind, bumps and nudges.
Nerves under the skin reposition slightly. You wait
for morning birds and perhaps a small hunger.
Perhaps you will eat.

Florida, A Season

Yesterday a rope of snake
threaded itself through the fence

waved its dominion
wrote its inscription.

Today I find its skin in the flower bed,
calamint blooming, sure sign of October

when palms let down their brown hair
and blades of the lily lie down.

I listen through diminished lush
to a changing resonance of bird calls.

One might easily miss this moment
under the peeling skin of summer

close to the quick as oranges swell
and absence, its hard seed buried

like a bone in deepening mulch.

After a season of empty branches

I walk every morning to see how many
vultures have returned to the pines.
First, two. Then half a dozen.

This morning I counted twenty-one.
A few had flown from the dense
needles to bones of a dead tree.

There they spread their wings like—
no, not angels—like scavenging birds
preparing wings for high spirals

that take them—no, not to heaven—
to work, important work to which they
assign themselves.

Once I sat at the end of a boardwalk.
Early. Alone. I sat writing.
One by one they came.

The first landed on an opposite rail,
the next on a rail behind me. Again
and again the black wings came

landed quietly near me, until I was
inside the flock, keeping motions
small, a pen moving on a page

my winged heart trying to ground itself.
When one lifted from behind me, her great
wing brushed my hair—no, not the spirit

of my beloved dead. Not Mother.

My Mother, Learning

i

She brought home leaves and stems,
pressed them between newspapers.
At night she took them out
spread them at the edges of the living room,
walked the circuit again and again, naming.
I lay in the dark, listened to her Latin chant.

ii

This is the book she took with her in the boat.
I turn the pages slowly, some wrinkled
from humidity, look at tinted plates
like watercolors fading—
anemones, sponges, crustacea, bivalves,
creatures in tubes under the mudflats—
images hidden in pages with long lists
of names. One by one she learned them.

iii

When the mosquitoes that did not bite
but ate larvae of the ones that did,
would not breed in the laboratory,
she brought them home, hung damp rags
from the ceiling of her bedroom,
turned them loose. For weeks, great care
not to leave the bedroom door open.

iv

Before she could afford to go to college
she stood in the factory ironing neckties
one after another, conjugating French verbs.

v

I am learning waterways, teaching my paddle
how to go without a chart. I rehearse the sequence
of channels and bays—Crooked Creek, Sunday Bay,
Oyster Bay, Alligator Bay…. Where all the mangroves
look the same, I stare until I see their difference.

One day, turning from Roberts River into
the cutoff, my arms are suddenly not mine.
I watch and feel my mother's arms, young and strong,
make every confident stroke, complete the turn.

Wildflower Pilgrim in the Rock Pinelands

I walk with a small leather book
open in my hands   chant   Latin
in rhythm with the body's motion

Pluchea rosea
Sabatia stellaris
Ruellia succulenta
the flower
of being unfolds

Melanthera parvifolia
Rhynchospora colorata
Slash pines quiver
in the wind's flowered breath

Centrosema virginianum
Heliotropium polyphyllum

Erigeron quercifolius
Crotolaria pumilla

I march through blossoms
pinwheels and chalices
stars and fireworks
journey everlasting

Bidens alba
Piriqueta
Coreopsis
Chamaecrista

With blossoms abundant
and waving fans
the pinelands bless
ancient and new

a passing pilgrim
gratitude blooming
offering praise
forever.

Benediction with Cello

I can never keep them completely separate—
blessings   and the French *blessures.*

What's the difference anyway
                              blessing   wound.
Think stigmata.   Think holy blood.

Torn by storm   struck by lightning
marked by char   every   one   of these pines
shows signs   a deep gash
bark ripped   heartwood gouged.

Beetles move in   ants come   and then
the probing woodpeckers   as streams
of balming resin pour   and thick lips form
around the busy wound   glowing   red

growing hard, its only purpose   healing
(unlike the simple clarity of sap
thin   conducting   daily business).

No wonder we   the ever-wounded   prize
the amber jewel's light   tremble   at tender
resonance   rosin   on the horsehair bow.

I I

The Naturalist

I went looking for alligators
but in the slender wetland reeds
a blood-red scarlet skimmer caught my eye.
I studied dragonflies all morning.

I went looking for dragonflies
but the red-shouldered hawk landed in cypress
fluffed his feathers, stared at me with serious eyes.
All afternoon we contemplated each other.

I went looking for hawks. Looking up,
I nearly stepped on the lubber grasshoppers,
females crossing the path with males on their backs—
places to go, lilies to eat. How could I not follow?

Today I'm not looking for anything.
I've learned.   Just go.

These *things with feathers*

　　　these gangly,
awkward,　squawky　things
these Great　Blue　　Herons　soon
will　　　fledge.

Now they fill the nest with
　　　legs and bills
that don't yet know
what to do　　　and silver guano,

make loud demands of their parents;
mouths gaping　　　　　　they shove
and fall
on each other's angles.

In the water below their pond apple tree
an alligator waits　　　alert
with great　blue　　hunger.

I watch the alligator
watching these oblivious birds.
Some will lose their balance　　　he knows.

I know　　how the world works　　　world
full of hungers that must be satisfied.
But I'm wanting a nest of　　　gangly hope
wanting these fuzzy angles

　　　　　　　　　all of them
to lift into a cloudless blue　　　and fly.

Alligator Mating

tender really
she initiates

a bit of stroking
a bit of caressing

       their deep throated song
       his water dance

if she likes him
if she allows it

       they swim, his arm
       on her back

if she likes this
if she allows it

       they glide together
       submerge together

like something
I've always wanted.

Alligator Evolution

like birds, they have gizzards
and complex hearts

cog-tooth valves
foramen of panizza

a mother constructs
her intricate nest

lays the clutch
in a circular pattern

tends the eggs
carries each hatchling

in careful teeth
to the water's edge

hunting, she leaps
to seize the heron

observe her legs
how they want to fly

her brain is the size of a walnut
it's all she needs

that walnut rides
at the back of my neck

recoils at the shadow
urges me to fly

and when I sleep
stays awake

dreaming, swimming
in an ancient sea.

the alligator's heart

i'm getting braver
        going out farther, deeper   alone
            reminding myself of the alligator's
                well-developed sense of hearing

repeating to myself
        what i say to others
            a gator in its natural state
                does not see a human as prey

my inclination in the wild
        move quietly   do not disturb
            but in the alligator's medium
                my heart   flutters

i slosh hard, bonk my stick on limestone
        give fair warning—*human coming*
            something big   *withdraw, withdraw*
                and of course he does—and would

even without my exaggerations
        so finely attuned he is to motion
            he knows i'm here
                but my own reptilian brain is at work

body alert beyond the frontal cortex's reason
        as i follow this trail made by an ancient
            know that somewhere near me beats
                his own four-chambered heart

Remarkable Ibis

Consider the ibis, remarkable bird,
its remarkable beak, remarkably curved.
In water stirred by remarkable feet,
it probes the mud for crustaceans to eat.

Do you know its remarkable family name?
Threskiornithidae—say that again.
Threskiornithidae: *sacred bird*, named
by the Greeks (what a wonderful word).

Ibis fly over in lines or in V's
flap-glide, flap-glide with remarkable ease
to marshes and mudflats, golf courses and lawns,
searching for crayfish, frogs, and prawns.

They're remarkably sexy in mating season,
their faces and beaks and even their legs
blush remarkable red as they cluck and coo,
rubbing bills with affection—eggs coming soon.

The ibis, though truly a bird of distinction
admired by Egyptians, engraved on their tombs,
has nevertheless escaped hats and extinction,
lucky for smooth un-remarkable plumes.

But even the famous John James Audubon
is known to have with pleasure fed upon
Chokoloskee chicken, a name we've heard
for the inimitable ibis, remarkable bird.

Mud Dauber

Back and forth between the   mud flat
and the underside of the   chickee roof
a solitary dauber's intense   industry

thread-waisted
golden-winged
singing

she pastes the sticky, granular   mud
forms a multi-chambered   nest
repeating   crescents

she flies, hunts
her blue-black body
full of paralytic sting

one by   one
she stuffs   spiders
into the   chambers

lays her   eggs
in the food   basket

Leave her alone.
She is busy.
She is not interested in you.

Nectar

Flowers here are diminutive.
I have to search for the small spot
of orange-red or lavender

flowers just big enough to attract
those who need attracting—the bee,
the wasp, the ant, the hummingbird.

These flowers waste no energy on me.
They don't need to catch my eye,
but I take the lesson, spend my energy

on the thing that matters, feeling
the pressure of sap in the stem,
making something small that blooms

for a solitary attention. A butterfly seizes
with its legs the Spanish-needle flower,
slowly opens and closes its wings,
balancing in air, sucking nectar.

Frog Music in July

In this damp little house at the swamp's edge,
I'm making a slow way to bed, having tea,
not wanting to close the door

on this wild-singing swampness,
contrapuntal frogness, an underlying
complex percussion of clicks

the roar and racket of vibrating
air sacs, tremulous membranes,
resonating mouth caves

trills, whistles, buzzes, snores,
brass bells, bassoons, flutes,
taut strings, triangles, rattles,

and every one of those mouths closed.
The whole urgent frog body swells,
vibrates, beckons to the beloved.

I think I will leave the door open,
swim in frog music till I fall asleep,
let its fervid longing fill
these shallow swamps, my ears.

Driving Loop Road

after cocoplums with their dark fruit,
wax myrtle, firebush, wild coffee

small openings, like keyholes
through which I could see

how a swamp darkens beyond fern
how a prairie extends into light

a young alligator sprawled in the road
three hawks held to their branches

shapes ahead of me scurried into scrub
an otter crossed in the rear view mirror

time was longer than it was,
so much in it—

the limestone gravel road
always narrowing

then the rain and milk-white puddles
wet green   and solitude

hawk time, alligator time
storm coming, rainy season

but since you ask, three and a half hours
dragonflies whirling over the road

Put a frog's tongue on my head

and you will hear me talk in my sleep.
I will tell you all my foolish secrets,
confess all night under that tongue

but I will not tell you
where the ghost orchid blooms
or where you'll find the pink tip

and green stripes of the tree snail. Some
secrets are sacred beyond your magic.
These I will keep until you have learned

to love the frog from whom
you have cut this tongue.

III

On a day when I have had hard news

I lie down and listen to the tender voice
of thunder—familiar, intimate.
It knew me in a time I can't remember

but my mother told me how she stood
in the edge of a great storm, held me.
I grew wide-eyed in her arms, *rapt*

she said, in wind-rush and leaf-tremble.
I knew then, we would share a love of this.

Now I am weary, heartsick. My mother is not here
to comfort. I lie down with the whisper of distant
thunder, hope it will come near—my dear
familiar thunder, its song and its caress.

Walking in the Neighborhood

It's not what I want.
Not a path through sawgrass
or a trail through slash pines.
Not where I could paddle a canoe.

It's what I've got
most ordinary mornings, but sometimes
I look up at a V of ibis flying toward
some vibrant, luscious mudflat.

This morning a little blue heron
perched on a wire—always surprising,
a large wild bird like that, balanced
on a line like a house sparrow.

She preened, then lifted her head,
stood still, as though her feet had settled
in soft mud at a stream edge
where she waits for a flicker.

I even look forward to those
aluminum sandhill cranes
at the house on Tinker Street,
concrete seahorses on a wall

and that purple gate in the high green fence,
its little sun burning, its yellow-flowering vines
leaping   out of control   taking over
growing wild.

Creaturely

Well before dawn I walk past houses
with dark windows, think of the humans
curled in their nests. They rest.

Soon the sun will come. They will rise
to resume collecting nuts, berries, slabs
of meat to provide for themselves
and their young. I walk

carrying a large palm frond
picked up from an empty lot
to use as underlayment for mulch.
An early car, the only one, passes,
puts me in headlights. I wonder

what the driver wonders
about this woman who carries
a palm frond in the dark.
A weapon perhaps? But how?

Strange the habits of humans
prowling with weapons,
sleeping oblivious,
putting each other in headlights,
passing each other in the dark.

Great Egret

Each day she steps a little closer to the window, extends

her long white neck, peers in at me. I practice

my casual-fellow-creature look, eyes resting on her briefly,

then scanning the yard as though my breath weren't

full of grace at her proximity, her whiteness,

that one wisp of wild feather rising. Today she watches me

feed my self – jaws opening – she understands this –

the grasping—the watchful eye—the gulp.

Learning from Dolphins

Startled at the first glimpse,
I turned to see them
tracing a quiet way

then followed
their benedictions

my pace matching
the pace of one
slow swimmer.

Each time it rose,
it rose beside me
until it disappeared.

I stand now and stare
at water, only water

learn what I must know of poems,
how they break the surface when least
expected, make their arcs in peripheral vision.

Apparitions cannot be commanded.
The air grows chill with pink consolation.
One line of light lies on the nervous surface.

Bee Lessons

In these early summer mornings
when the wild coffee's flower clusters
release a scent sweet as jasmine
that draws the bees, makes them mad
with joy and striving, I lean into bloom
crazy as the bees, avoiding sting

but they are oblivious to me,
focused on the morning's meaning,
teaching me the sweet of need,
the buzz and whorl of making.

Easter

For you, this day has meaning. I must construct it,
wonder what is this thing called *god*, so persistent
in the human psyche, catalyst for wars and beauty.

I hear you say that *god* is within us. Rather, are we in *god*,
molecules or parasites in some great rushing bloodstream,
elements of a vast body, dying and replenishing—eternally?

All of history's philosophy has not resolved such things.
Science contradicts and discusses with itself—or acknowledges
the question is beyond its repertoire.
                    Well, it's Easter.
Let's go to the beach in early dark, try again to transcend grief
by ocean, wind and (non)belief. We will be whatever it is we are,
listen to the medium that contains us, as it rushes and roars, beats
like a pulse, strange and terrible, making beauty. Full of ignorance
and sunrise, we will stand on the shifting, shell-littered shore.

I lie down in darkness beside the window

with the shadow-presence of tall pines
and the wild coffee's variegated leaves.

I know the purpling cocoplums are there
and sea grape clusters preparing to ripen.

All of these I planted. I have planted
my own peace. A canopy of gumbo limbo

gathers above its photosynthetic bark
waits for light's alchemical transformations.

Lemon tree, banana palm, blackberry vine
attend the window where I lie down

with my restlessness, my rootlessness, my wants.
Trees embrace feathers of the sleeping hawk,

the waking owl. *Who-whoo-whoo-whoo.* I listen,
watch for the moon's rising, know exactly
where it will slide into branches.

IV

Moonflower
Ipomoea alba

In a field spun with grasses and milkweed vine,
folded *Ipomoea alba* blossoms stand erect
above a green perennial tide, wait for moon time.

With or without a moon they will bloom,
open wide their tender throats,
fluted funnels, alabaster white

extending pale green nectar guides.
In the way of flowers, they long for the touch,
the tremble, a sphinx moth's flutter.

Their entire being is made for this, soft
trumpets emitting scent—honeyed, complex,
maddening, irresistible, a silent siren's call

from lush radially symmetrical displays
of stamens, sepals (starfish, daisy),
one distinct pistil. An evening wind

ruffles blossoms and heart-shaped leaves,
spreads the invitation. Moths come
dizzy with scent and desire, nuzzling.

We saw moonflowers at Watsons Place
when we camped there. They bloomed
the night of the alligator. A night's hungers

are sometimes quiet, sometimes crashing
like ancient jaws—in the morning, flowers
droop beside the freshly broken claws.

Heart Lessons

The snail carries its two-chambered pump in a heart bag.

The earthworm's heart is a flood in five stages.

The cockroach has a dozen hearts. No matter if one falters.

When the octopus swims, all three hearts stop beating.

The jeweled hummingbird's heart is perpetually thrilled.

Emperor penguins' hearts are zen. Slow.

In winter, the wood frog's body freezes, the heart stops. Till spring.

In the glass frog's translucent body, watch the bright heart beating.

The beautiful sea star has no heart, sets my heart racing.

If the zebrafish has a wounded heart, it grows a new one.

Lichens this morning

the fungi enclosed the algae within their tissues
in a most intimate embrace.
                              --David Attenborough

Lichens abound here, on almost every trunk
splotches of grey-white, grey-green, yellow, red—
a pucker here, a ruffle there, growing

one half inch per year. Their development can't be
forced. I reach out, tentative, almost touch
three different textures—*crustose, foliose, fructiose.*

It's the rainy season now. I want to see them again
in a dry season, and so I begin to imagine a future for us.

Looking into cypress, rough trunks freckled with lichen,
I remember when I believed you and I might grow
lichenous, better together than apart.

I wade alone in the littoral zone with wild aquatic
flowers, step towards the wet amorous world
of lichen, wormvine, pimpernel, butterfly orchid,

watch from the corner of my eye
one slow alligator on the prowl.

Love Vine
Cassytha filiformis

It drapes its bright net on the skeletal form
of the now unrecognizable blackened bush
a tango of golden strands curling

sinuous—at first, a single wild wisp reaching,
then… the grasp. Crush it: sassafras, cinnamon.

Whoever named this knew something
how the strands creep, adhere, loop
then penetrate intimate barriers.

Membrane to membrane this vine
withdraws what it needs

in a tangle of clinging, matted stems—
aga-mula-neti-wel, plant with no beginning
or end. It's said in the islands that if

you want to make someone love you
brew the vine with gumbo limbo,
a fine tea with or without cream.

If you need to rid the head of lice,
apply the sap.

A Note on Abandonment
Bursera simaruba

My fascination with the gumbo limbo
has nothing to do with its red
heartwood finely carved
into carousel horses and ostriches

nothing to do with the efficacy
of teas and aromatic baths
or the way light streams
down a muscular photosynthetic trunk,

trickles on bright exfoliating bark
though I'm getting closer to the heart
of the matter, how a sturdy fence post
makes new branches, spreads new leaves

how every lopped-off limb retains
suck-sap, light-slurp, leaf-leap—
an irrepressible urge to flourish.

Outside my window cardinals perch
and preen in this determined tree

that grew

from a stick I stuck in the ground
and left alone.

V

Resisting the Poetry Exercise

*--Put yourself back
in the house you grew up in.*

No. This house is square and white and fine.
Camellias and azaleas announce the yard.
Evenings I enter the high heavy door,
the grandfather clock strikes five.

I don't want to go back to that small house
those rooms, furnished with found things,
chairs that washed up in storms,
rickety dressers with drawers that slid easily
neither out nor in. But wait--
there was a bookcase.

Paradise Lost in a pale blue cover,
Complete Works of Shakespeare with real
gold on the spine, and Mama's big
biology book, worn from use and hope
and waiting, the one I sat and tore and tore.
It was the only time I saw my Mama cry.

Homecoming Aubade

Pelicans fly in V formation low over the waves
in which, as a child, I swam. Ocean into which my father
sank with a shrimp boat, nearly died
dear ocean, treacherous ocean, home ocean

where my uncles fished and during the war, sailed
with the merchant marine, where my mother learned
the name of every creature that swam or burrowed in mud
dear ocean, treacherous ocean, home ocean

I crossed it when I first left home, naive and eager.
When my brother died, these are the waves I wanted to ride
rock me, rock me   tide of grief
                    dear ocean, treacherous ocean, home ocean

Now this purple morning alive with wings and the arcs
of dolphins, glows   red veins, yellow lasers   slowly
come the ordinary blues of day
dear ocean, treacherous ocean, home ocean

One jagged stain of coral almost-red
persists above the horizon like a wound.

The prediction for today: sunny and cold.

Timex: A History

Five minutes into the swim I realized
I was wearing my watch

remembered my father with a jeweler's loop
bending over his workbench with mysterious
small bits he had taken out of a watch,
would put back in. He hated Timex. *Junk.*
No jewels. Not a ruby or diamond or anything
that would last.

I took off the watch, set it on the edge of the pool

remembered the old tv advertisements, black and white,
Timex strapped to the propeller of an outboard motor,
the whirl, and then….*Timex takes a licking
and keeps on ticking.*

Three days later I look at my drenched watch.
It reads 3:11. Accurate.

My father never saw a digital watch or one called
water resistant. What now would be his assessment?
What would he say about this age
in which nothing is expected to last

he who died at 35. No rubies. No diamonds.

Collision

Father, we were like those lanterns
I used to watch on the water at night,
silent lights of invisible fishermen
moving separately at great distances
each a mystery to the other but moving
through the same darkness, the same water,

and so—that night when I ran to tell you
that if you went out again, Mama
who was already packing your suitcase
would put you on a bus to Raleigh, when

breathless I caught up with you, and you
asked why and I flung my bony body against
your bare chest, sobbed, *because you're going
to get drunk again*

when you wondered if I wanted you to stay
and I said yes, when you held me
and did not go—we reeled
at our single gentle collision in the dark.

Feathers and Song of the Dove

I watch the doves, fluttery, flirty,
ruffling feathers, murmuring
in the morning yard,
mourning yard,
dovie, dovie, Dovie

who told the story of a well-patched dress
and the yellow cotton mittens her mother made—
picked the cotton, carded and spun, yellow dye
from a plant whose name I've forgotten.

Dovie, Dovie country-come-to-town
became a nurse in a crisp cotton uniform,
nursed the wife devotedly,
then married the widower
who would make her a widow.

Grandmother Dovie who held me,
rocked me, sang a lullaby
of her own invention,
borrowing the tune of a hymn.
  Bye-oh-bye my precious baby
  Go to sleep and don't you cry.
Doves of the yard coo,
shift between underbrush and tree,
flickering in and out of sight,
soft shades of white and gray.

The Little Back Bedroom with a Tin Roof

is where my grandmother liked to sleep
hoping for the soft percussion of rain

or if not rain then wind to bend the branches,
make them brush the tin edge,

gentle soporific rackets with rhythms
to ease her into sleep, sing away

the ache of her body which knew more
of work than a woman's body should,

husband in the grave mid-Depression,
her beloved home place burned to ash.

Sometimes on afternoons that promised rain
she'd say, *Come, Sugar Plum,*
lie down with me. Let's listen to the rain.

Watching the Four O'clocks Close
Mirabilis jalapa

Wild now, creeping from the tangle
of an undeveloped lot, escapees
from domestication, these four o'clocks
startle open a pathway into the forgotten.

Every evening when Aunt Mary called
It's four o'clock time! we bounded out
the screen door, Aunt Mary drying her hands
on her apron, letting the door slam behind us.

Then we stood at the chimney garden
as flowers folded in slow pink synchrony
the bush of bloom transforming into a bush
of buds, the promise of tomorrow.

Observing the quiet drama
always predictable, always new,
we never spoke.

What have I lost of slow,
standing quietly at a chimney
witnessing cycles of flourish and fade?

When flowers had closed, almost dinner time,
we went back into the house, light fading
from weathered boards and the dirt-swept yard.

Bones of Ancestors

Oh, so many beloved bones,
bones of my ancestors,
how could I count them

bones held together by lace
bunched in a cluster
between the shoulder blades

bones stripped
by the shark's teeth

bones that loved bones
of piano—a music
intricate now in stillness

bones in the ashes
of a burned house.

I loved my father's uneven bones,
one leg shorter than the other—a limp.
"Tall and lanky" they said he was.

I learned the word—my daddy
was lanky—clung to the long leg,
rode the swinging bone.

Bone slowly
dissolving.

Inside their flesh, I feel my bones
secure, upright, bearing weight,
anticipating motion, bones intrepid,
bones of somebody's ancestor.

Invocation with Red Sails

Hōkūle'a, teach me how to be on the dark sea
without a chart, clouds obscuring stars.
Teach me how to hold back panic, read the waves.
Teach me to trust the ancestors, who knew more
than I yet know how to know.
I am on the sea now, learning,
making my way where there is no path.
I navigate through terror, seek direction
from each swell of the sea, attention absolute,
destination invisible but I know it's there.
Others before me have made this journey
in a simple craft. I must make it, too, believing
in a moment when, beyond this tumultuous passage,
I will see a quiet beach, a shining spit of sand.

V I

Red Mangrove
Rhizophora mangle

It stands in salt but cannot drink it
a story of molecular ingenuity
clever tissues, no taproot

and yet this tree is mostly root.
Aerials leap from high branches,
rooty guy-ropes dangle, sway

touch bottom, dig, rise again
loop along and interlace,
transfigure into flying buttresses

hold through storm,
create a kind of grace.

Red mangrove, I come in pilgrimage,
paddling my cumbersome canoe

seeking pollination by wind,
holding up my small, simple
almost invisible flower.

How to Spend the Night in a Canoe, Coldest Night of the Year

When a north wind has blown the water from around your boat,
left you surrounded by mud deep enough to swallow you whole,
when you have taken the shock of a rising tide that stopped short,
turned away   acknowledge the inevitable.   Get to work.

Remove everything that has served its purpose—empty canisters,
empty jugs; set them on mud. Spread tarps, mats, sleeping bags
in the bottom of the canoe. Put on all the clothes you can and use
the rest to insulate the boat. Lie down in the nest you have made,
pull over you the folded tent.   Wait.

The moon you have watched grow larger each night, nearly full,
floats with Orion, balances over the wooden paddle planted in mud.
Trace with your eyes the dark shapes of unnamed mangrove islands.
Listen for the dolphin's blow in the channel you can't quite get to.
Feel how your face grows colder   your body warmer.   This

is wilderness. It holds you, pins you   at the center of the universe,
suspends you in the essential. Time has nothing to do with clocks.
Tides defy their charts.   Watch

the slow arcs of passage   stars and moon. Feel yourself
warm   free   breathing.   Know
how lucky you are.

Anchoring

I paddle through saline water deep
into the forest, the muddy carbon sink,
inhale. Oxygen in my lungs is deep
peace. Breathe. Stroke. Breathe.

In this green fringe, storm buffer
for the land, my own storms begin
to quiet. The howl subsides.
Breathe. Stroke. Breathe. Begin.

Water parts with quiet ripples.
Tide falls, leaves a dark band
like memory, on arching roots that grip
an invisible substrate. In bottomland

I am new, a propagule that floats
until she lodges somewhere, an accident
of landing. I reach into what I cannot see,
feel and grasp the thing that anchors me.

Island Poem with Recollections of Frost and Bishop

Paddling today through water mazes   fragments   flickered
memory flashed glinted in rhythmic motions   tides of then
and now   tricky navigations—*two channels diverged.*

A sturdy rope secures the canoe to this island's one Black Mangrove.
I am   so far   from anywhere   dependent on this small craft
remembering the knife that Crusoe *implored not to break*
                    how it *reeked with meaning.*

I press my feet into a ragged pile of flotsam   fragments of fan shell
broken worm tubes, crab carapaces, sponge torn from the living reef,
oyster shells   spots of lavender   color of bruise where muscle

tore free.   In this beautiful   sea-beaten   mollusk debris,
violence   and one perfect gastropod   glassy orange at the orifice.

Ibis in a Mangrove Sunset

These ibis who work so hard all day
in marl prairies and on the mudflats,
thrusting beaks into soft sediments,
probing for prey

like to loaf in the evening,
in bone branches of old buttonwood
where in late sun they gather, lit
like ancient torches and candles.

They do not trouble each other with gossip,
having clucked among themselves all day
in shallow pools and muddy fields of crustacea.

Soon one will signal; they will rise,
fly to the green-black mangrove island,
into the hushed and placid night
a thousand white-winged susurrations.

Green Hell I've heard it called

where green grows out of mud and gray-green water
makes a tangle   reaches out for me   snatches

bungees from the tarp   scrapes things out of the canoe
points me forward   grabs   and holds me back
entices me into its   green   teasing   curl

shakes in wind   flips up salty petticoats   Green
does not hate me   or love me   or guide me   or lose me
I guide or lose myself   and green is there   a witness.

In these ragged tunnels   green is necessity   Explain it
with the physics of light   wavelength 495-570 nanometers
it makes no difference to know all that   Not envy

not spleen   not any of the things they say   not metaphor.
I grew up with green everywhere   green storm
green tide   green rising from backyard dirt

I took it for granted   Green took up residence in me
transformed itself into need   My first canoe was green
My now canoe   color of senescent leaves

I have never said   green   was my favorite color
or malachite   shamrock   olive   emerald   jade
Green is the color of loss   Caedmon wrote that Adam

stepped on *grene graes*   Green is in the eye
of the beholder optic   retinal   occipital
I lift mine eyes unto *Grene*   sufficient *Grene*   living

Grene   transcendent   trembling   need   in wind.

Navigation

Here in the uncharted dark   canoe tied to buttonwood and mangrove
we float between the deep channel   and a high muddy bank.

Not hurrying sleep   though I am tired from searching in the dark
I look up from my floating cradle   into arches of mangrove root.

Moonlit, they rise from mud   then descend like benediction
almost touch the boat. I remember arches of Notre Dame and Chartres

from which I've navigated far   serenity   grandeur of forms
resonance   of something old   necessary   un-nameable.

Twice in the night I awaken   once hearing the dolphins' breath
once startled by the big dipper   poised to lift water from our channel.

In the morning I see   shining   accurate   directly ahead of us
the marker we navigated toward in darkness   could not find.

VII

Continuance

I am thinking of Earl who shot himself last night.
I am thinking of Sue who will not be here in the fall

and all those ancestors lining up in genealogical charts
where a space waits for a grandchild to write my name.

I am thinking of the thousands of earth-stained skulls
heaped in catacombs below urgent Paris traffic

and nameless bones dissolving in sea or rising
as smoke into air, bones disarticulating in dirt.

We are all temporary   and particulate.
We are all particulate   and permanent.

Our name is earth. Our name is sea. Our name is air.
We quake and storm, turn on an axis, revolve around the sun.

Lacking my mother's arms

I reach
for the green   embrace

of a wild forest
that stands   in water

as did she
filled with   wonder.

Into the darkening
forest   I sing

croak, burble, screech, caw,
keen into frog-throb,
wind-wail, thunder

rocked in wilderness roar.

Return to Water

I was ten when my father's boat went down.
Two days we didn't know where he was.
Then he lay breathing on the sand.

But mostly water was simple –
high, low, high, low. We investigated
sloughs and marshes, seeking treasure:
crab, seahorse, moonsnail, anemone.

I've come back to live with tides,
their gelatinous and calcarious forms.
I've learned. This proximity matters.

For years the nightmare repeated–
a tidal wave rose, I couldn't run fast enough.

Last night again I dreamed a mountain of water rising.
This time I wasn't on the shore, I was in the sea
thrust upward in a giant swell, rising

dreading the crash, the dashing of my bones
abrasion of shell, skull cracking.
But at the shore the wave lay down,
set me on a quiet sand.

Because of the Mudflat I Bought the House

Ghost shrimp flick wildly in a tidepool,
crowned conchs slide along sinuous trails,
and that living fossil, the horseshoe crab,
makes its intricate, many-footed inscription.

Weary with sentences, paragraphs, pages
lining up in academic order, I put on the old
shoes, walked out the door and down the road,
stepped onto muddy gel and ooze, felt its give
as it took my weight, offering up a familiar
sulphureous smell—sweet
to one who grew up with it.

Spider crabs, banded tulips, a hermit crab
in a moon snail shell, great heart cockles,
twirling strands of conch eggs.

As long ago my mother did, I lean and look
and name aloud, each name a surprise
in my mouth where it has lived in hiding.

In mud and water, carapaces and light,
amid rhythms of an intertidal world
I find   not one   straight line.

Anemone

What part of me would move like this
if taken from me, returned to the sea,
exact equivalents of salt—pancreas or liver
perhaps, attached to a calcium substrate,
waving, filtering, expanding and contracting
in the garden of sting—
small fish nudging at soft edges.

It's the invertebrates I feel closest to,
protoplasm and saline our common origin.
Such clarity—the moon jelly's light and pulse.

Awkward these appendages of mine. But once
they held a fluid motion, longer ago than I
can imagine. Genes remember when I was
something older than consciousness,
more beautiful than any human construct.

We are tender, my friends. We are octopus and squid,
water and salt, a bit of jelly complicated by the history
of bone. Close your eyes. Feel your body anchored
and waving, all of you, all of you, pulse and feeding.

Looking for Tetrazygia

The sky darkens as I drive slowly, look through rain
for blooms withheld in that easy Everglades dry season
of blue skies, perfect weather. I park in front of a sign
Closed for the Season stare through rain-blur into the pineland

        where years ago I came, wracked with pain,
        glowing with joy, something awakening
        from ashes of grief—alive, wild, full of leaves
        in strange configurations. Tetrazygia's
        patterned leaves woke me from a daze of ache.

Rain intensifies, slams against the windows, thunder rolls
and suddenly I want to weep, rain and weeping
the waters that make bloom in a violent clamor.

I consider flouting lightning, walking in search
of elusive flowers. What better way to die
than searching for blooms of West Indian Lilac in July?

*

I see it then, a shining cluster of white bloom
in the dark-wet animated leaves whose parallel
venation speaks of ancient lineage.

This rain is fierce with the force of a planet
green and blue with becoming—flowering,
fading, seeding. I should get out, stand committed
in the same rain as these deeply veined leaves,

this wet and glowing flower. When I was young,
I would have. Older now I am less dramatic
in my solidarity, but I am inhabited by a joy
more ancient than my particular protoplasm.

I cannot understand it. I do not need to.
I lift my eyes to the limber slash-pine forest,
its slow swaying in a wash of gray.

Pileated Woodpecker

Hello, your Oddness, Sir Fickleness leaping
pine to pine, flicking chunks of bark

Lord of the Redheads, Sir of the Topknot,
handsome bird in a striped tuxedo

granting me audience, showing off
your fine profile and pattern

then leaving with deep wingbeats,
drawing a line of your own design

reconfiguring marvelous air. I stand,
stare, hear your laughter, laugh back.

Poem with Wings

At the window I sit with patterns of light, a brush of wind,
absence in presence, a very small bird.

Peace and a restless spirit ride the same breath.
Loss and fulfillment walk in the same wind.

I am the tree, rooted and sturdy, I am the wind passing through.
I am the fruit and its nemesis worm.
I am dust, I am desert, my banks overflow with flood.
I am the carcass and the vulture circling higher.

Each day I sit with the same sun, the same wind. And not. And clouds.
And rain. Never a repeated plot. I stare at the days, one after another,
delights and miseries of leaf and weather, the come-and-go of wings.

If I could sit long enough….
If I could stare long enough…

One restlessness or another always presses. How can I know
what would happen if I stayed and stared, pleasure
and ache conjoined? How long is long enough?

And now the cardinals come, brilliant in their red perfection,
knowing what they are, claiming their place in the order
of the yard, open-beaked, full-throated, singing.

Thistle Song
Cirsium horridulum

What a wooly mess this horrible thistle is   spilling
itself out of the spent flower   purple explosion
attracting butterflies   bees   beetles   wasps   me.

I admire the silk-borne seed   finely sculpted leaves
thorns long and precise   wonder if I am
who I think I am   who I want to be.

In Greek, *kirkas*, swollen vein.
In Latin, *horidulum*, prickly.

Am I?

Between thumb and forefinger   I take a piece
like softly matted fur   speckled with elongated seed.

It's good out here at the center of a prairie   circled
by swaying slash pine   bent over *Cirsium horridulum*.

Horridulum. I love this name   how it opens my mouth
makes a roundness   seed fluff could fit inside.

I could hold it there   not closing   not swallowing
simply containing the feathery silk of seed   until

a hooked wind reaches in   lifts it out   spreads it wide
damp   with my moisture   the loving wet of me

until it settles into some fertile crevasse
roots   rises   explodes.

EPILOGUE

Notes from a Marine Biologist's Daughter

My mother loves the salty mud of estuaries,
has no need of charts to know what time
low tide will come. She lives
by an arithmetic of moon,
calculates emergences of mud,

waits for all that crawls there, lays eggs,
buries itself in the shallow edges
of streamlets and pools. She digs
for *chaetopterus*, yellow and orange
worms that look like lace.

She leads me where *renilla* bloom
purple and white colonial lives,
where brittle stars, like moss,
cling to stone. She knows
where the sea horse wraps its tail
and the unseen lives of plankton.

My mother walks and sinks into an ooze,
centuries of organisms ground
to pasty darkness. The sun
burns at her shoulders
in its slow passage across the sky.
Light waves like pincers
in her mud-dark hair.

ACKNOWLEDGMENTS

I offer grateful acknowledgment to the editors of publications in which some of these poems or their earlier versions appeared:

Cave Wall—Driving Loop Road

Cold Mountain Review—Benediction with Cello; Lexicon

Deep Wild—How to Spend the Night in a Canoe, Coldest Night of the Year

Getting the Knack (anthology)—Resisting the Poetry Assignment

Gettysburg Review—Notes from a Marine Biologist's Daughter

Hawaii Pacific Review—Invocation with Red Sails

Insects and Texts (anthology)—Mud Dauber

Poetry South—Collision

Saw Palm: Florida Literature and Art—Alligator Mating; Walking in the Neighborhood

Tar River Poetry—Bones of Ancestors

Twelve Mile Review—A Note on Abandonment

In one instance, first publication was in a different form:

O Miami / Everglades National Park Poetry Banner Project—Remarkable Ibis

Over time, this work has been supported by residencies in Everglades National Park, Big Cypress National Preserve, and South Carolina State Parks.

For the ongoing support, companionship, critique and encouragement of the Frondes—Annie, Babo, Deena, Kathy, and Pat—I will never have sufficient words to express my gratitude. The same must be said of the Warren Wilson Alumni Community. Thank you endlessly.

Thanks to Hugh and Elvie for their *spirit of aloha* and magical foods as I worked on this manuscript (and others) in the spirit-supporting Mango House.

Special gratitude to my family, including future generations. They inspire me.

And to Lynn, who dwells with me in possibility.

NOTES

These *things with feathers*—The italicized phrase references Emily Dickinson's poem "Hope is the thing with feathers."

Remarkable Ibis—Chokoloskee is an island at the northern edge of Everglades National Park. Island pioneers of the early twentieth century included ibis in their diet, a practice that gave rise to the appellation "Chokoloskee chicken."

Put a frog's tongue on my head—Somewhere in my extensive naturalist readings, I encountered the belief among peoples of a Caribbean island that putting a frog's tongue on a sleeper's forehead would cause them to speak their secrets. Though I have searched, I have not succeeded in locating the source.

Invocation with Red Sails—*Hōkūle'a* is a double-hulled sailing canoe, a replica of the Polynesian craft that first brought people to the island of Hawaii. In 1976, a mostly young crew of Hawaiians, successfully sailed from Hawaii to Tahiti and back using no navigational instruments, guided only by stars, waves, birds and natural phenomena.

Island Poem with Recollections of Frost and Bishop—The phrase *two channels diverged* echoes Robert Frost's famous "Two roads diverged" in the poem "The Road Not Taken." The phrases *implored not to break* and *reeked with meaning* quote from Elizabeth Bishop's "Crusoe in England."

Notes from a Marine Biologist's Daughter—This was first the title of a poem at *Gettysburg Review*. Then it was the title of an article which included the poem at *Harvard Educational Review*. Now it is the title of a book. Here, the title retires, but I will continue to think of all my work as notes from a marine biologist's daughter.

ABOUT THE AUTHOR

Anne McCrary Sullivan grew up on Wrightsville Sound in southeastern North Carolina. She taught high school and raised two sons in Texas, then had a second career in university teaching. Along the way, she earned an MFA in Poetry from the Warren Wilson MFA Program for Writers and a PhD in English Education from the University of Florida. In 2012-2013 she was a Fulbright Scholar in Calabar, Nigeria.

Living now on the Gulf Coast of Florida, she is a Florida Master Naturalist, wilderness canoeist, and writer. Her books include *Ecology II: Throat Song from the Everglades* and *Learning Calabar: Notes from a Poet's Year in Nigeria.* She is co-author with Holly Genzen of *Paddling the Everglades Wilderness Waterway* and *The Everglades: Stories of Grit and Spirit from the Mangrove Wilderness.*

She is a life member of the Florida Native Plant Society, the Gopher Tortoise Council, and the League of Environmental Educators of Florida.

Visit her Amazon author page at: *amazon.com/author/annemccrarysullivan.* Find her on the web at *www.annemccrarysullivan.com.* She rarely uses her Twitter account, but you can give her a try *@tetrazygia.* She is on Facebook.

CPSIA information can be obtained
at www.ICGtesting.com
Printed in the USA
BVHW021925040523
663609BV00008B/113

9 781955 194143